Practice

Eureka Math®
Grade K Fluency

Published by Great Minds®.

Printed in the U.S.A.
This book may be purchased from the publisher at eureka-math.org.
CMP 10 9 8 7 6 5 4 3 2 1

ISBN 978-1-64054-552-6

GK-M1-M6-P/F-04.2018

Learn • Practice • Succeed

Eureka Math® student materials for *A Story of Units®* (K–5) are available in the *Learn, Practice, Succeed* trio. This series supports differentiation and remediation while keeping student materials organized and accessible. Educators will find that the *Learn, Practice,* and *Succeed* series also offers coherent—and therefore, more effective—resources for Response to Intervention (RTI), extra practice, and summer learning.

Learn

Eureka Math Learn serves as a student's in-class companion where they show their thinking, share what they know, and watch their knowledge build every day. *Learn* assembles the daily classwork—Application Problems, Exit Tickets, Problem Sets, templates—in an easily stored and navigated volume.

Practice

Each *Eureka Math* lesson begins with a series of energetic, joyous fluency activities, including those found in *Eureka Math Practice.* Students who are fluent in their math facts can master more material more deeply. With *Practice,* students build competence in newly acquired skills and reinforce previous learning in preparation for the next lesson.

Together, *Learn* and *Practice* provide all the print materials students will use for their core math instruction.

Succeed

Eureka Math Succeed enables students to work individually toward mastery. These additional problem sets align lesson by lesson with classroom instruction, making them ideal for use as homework or extra practice. Each problem set is accompanied by a Homework Helper, a set of worked examples that illustrate how to solve similar problems.

Teachers and tutors can use *Succeed* books from prior grade levels as curriculum-consistent tools for filling gaps in foundational knowledge. Students will thrive and progress more quickly as familiar models facilitate connections to their current grade-level content.

Students, families, and educators:

Thank you for being part of the *Eureka Math*® community, where we celebrate the joy, wonder, and thrill of mathematics. One of the most obvious ways we display our excitement is through the fluency activities provided in *Eureka Math Practice*.

What is fluency in mathematics?

You may think of *fluency* as associated with the language arts, where it refers to speaking and writing with ease. In prekindergarten through grade 5, the *Eureka Math* curriculum contains multiple daily opportunities to build fluency *in mathematics*. Each is designed with the same notion—growing every student's ability to use mathematics *with ease*. Fluency experiences are generally fast-paced and energetic, celebrating improvement and focusing on recognizing patterns and connections within the material. They are not intended to be graded.

Eureka Math fluency activities provide differentiated practice through a variety of formats—some are conducted orally, some use manipulatives, others use a personal whiteboard, and still others use a handout and paper-and-pencil format. *Eureka Math Practice* provides each student with the printed fluency exercises for his or her grade level.

What is a Sprint?

Many printed fluency activities utilize the format we call a Sprint. These exercises build speed and accuracy with already acquired skills. Used when students are nearing optimum profi ciency, Sprints leverage tempo to build a low-stakes adrenaline boost that increases memory and recall. Their intentional design makes Sprints inherently differentiated; the problems build from simple to complex, with the first quadrant of problems being the simplest and each subsequent quadrant adding complexity. Further, intentional patterns within the sequence of problems engage students' higher order thinking skills.

The suggested format for delivering a Sprint calls for students to do two consecutive Sprints (labeled A and B) on the same skill, each timed at one minute. Students pause between Sprints to articulate the patterns they noticed as they worked the fi rst Sprint. Noticing the patterns often provides a natural boost to their performance on the second Sprint.

Sprints can be conducted with an untimed protocol as well. The untimed protocol is highly recommended when students are still building confi dence with the level of complexity of the fi rst quadrant of problems. Once all students are prepared for success on the Sprint, the work of improving speed and accuracy with the energy of a timed protocol is often welcome and invigorating.

Where can I find other fluency activities?

The *Eureka Math Teacher Edition* guides educators in the delivery of all fluency activities for each lesson, including those that do not require print materials. Additionally, the *Eureka Digital Suite* provides access to the fl uency activities for all grade levels, searchable by standard or lesson.

Best wishes for a year filled with aha moments!

Jill Diniz
Director of Mathematics
Great Minds

Contents

Module 5

Module 6

Grade K
Module 1

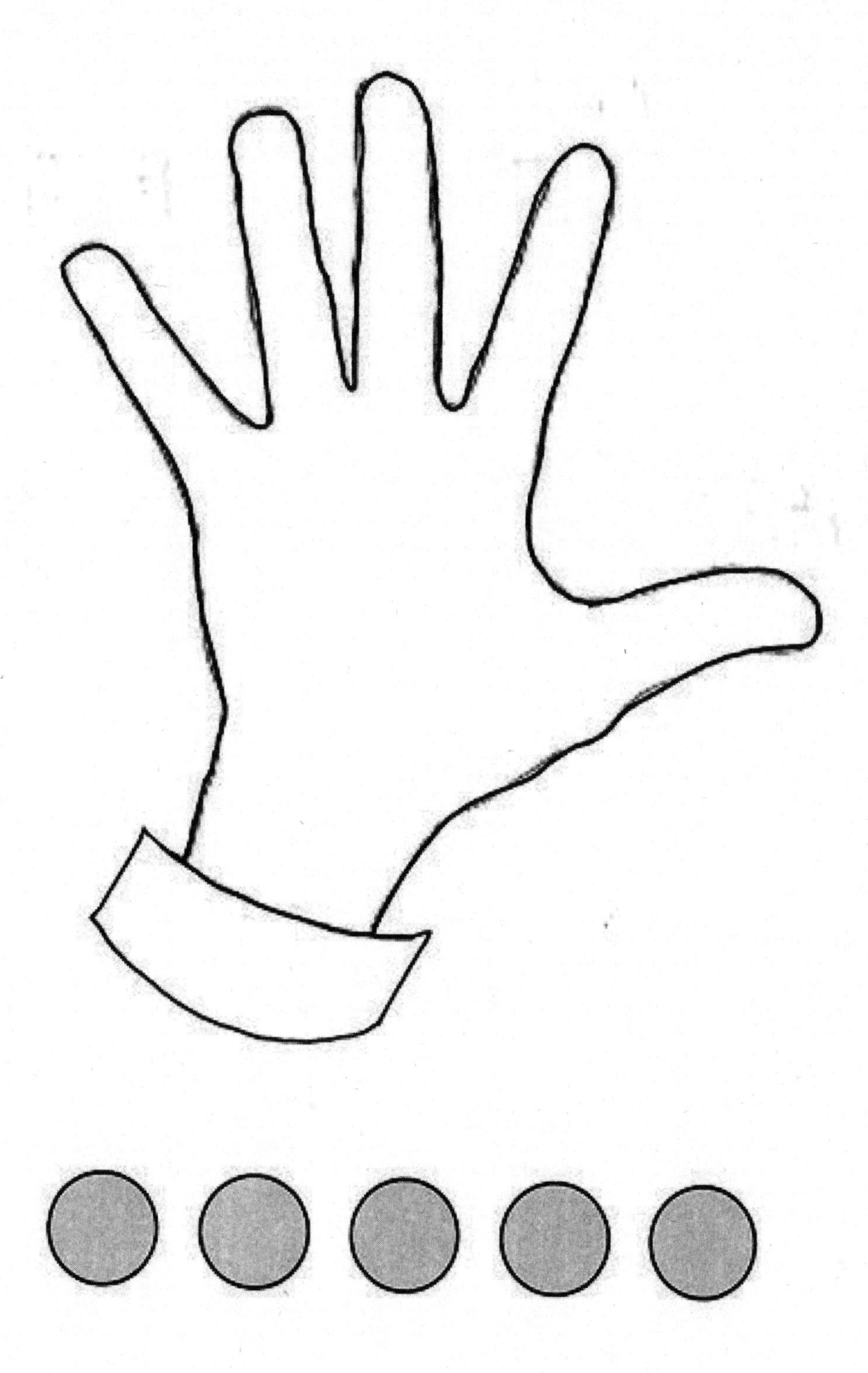

left hand mat

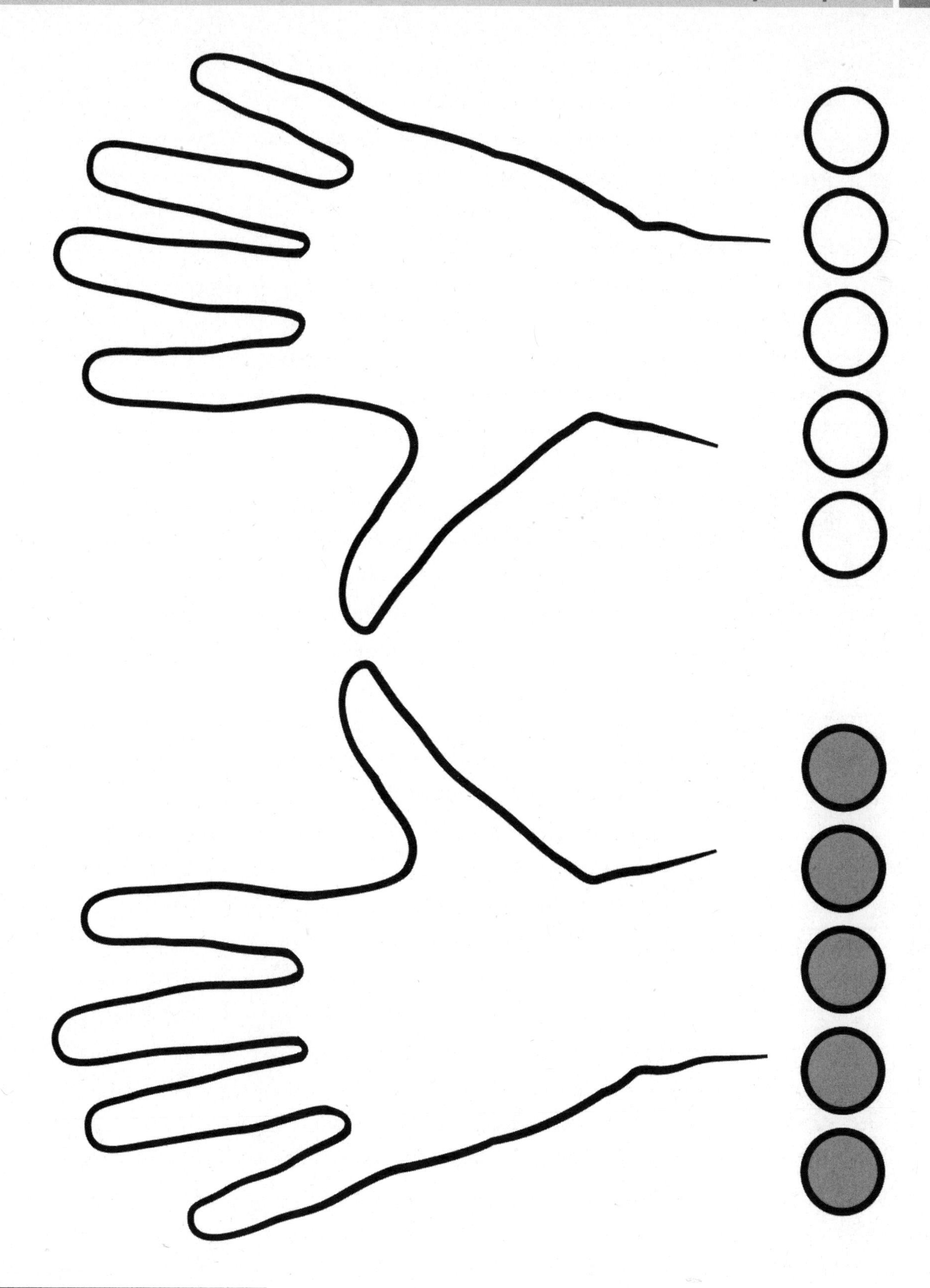

two hands mat

piggy bank mat

EUREKA MATH®

Lesson 29: Order and match numeral and dot cards from 1 to 10. State 1 more than a given number.

7

Name ______________________________ Date ____________

Draw 1 more, and write how many in the box.

	How many?		How many?

draw 1 more

Lesson 32: Arrange, analyze, and draw sequences of quantities of 1 more, beginning with numbers other than 1.

9

Name ____________________ Date __________

Draw 1 more, and write how many in the box.

	How many?		How many?

draw 1 more; from Lesson 32

Grade K
Module 2

Draw more to make 5.

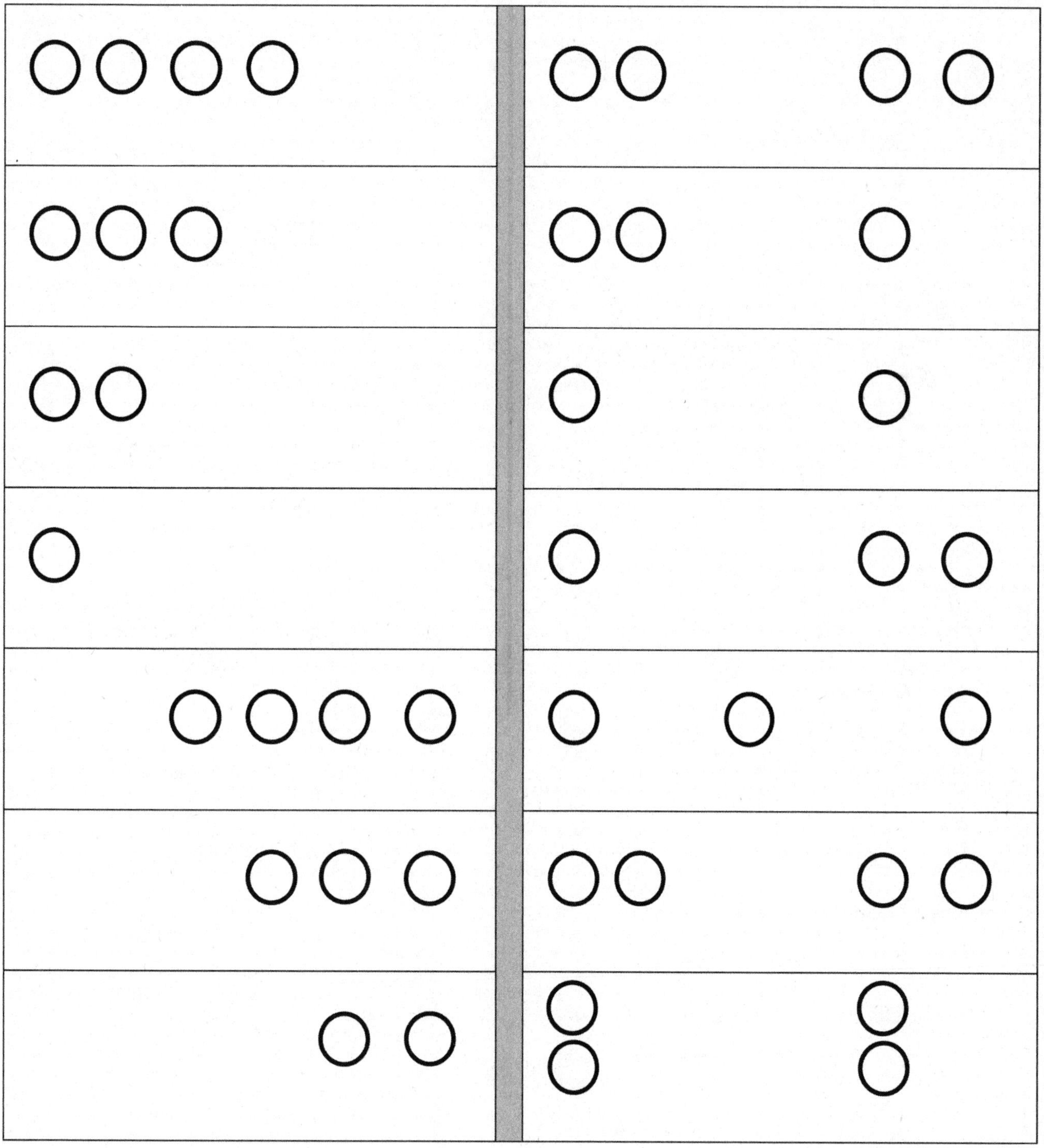

draw more

Grade K
Module 3

hidden numbers mat

Draw more objects, or cross out objects to make 5. Circle the group of 5.

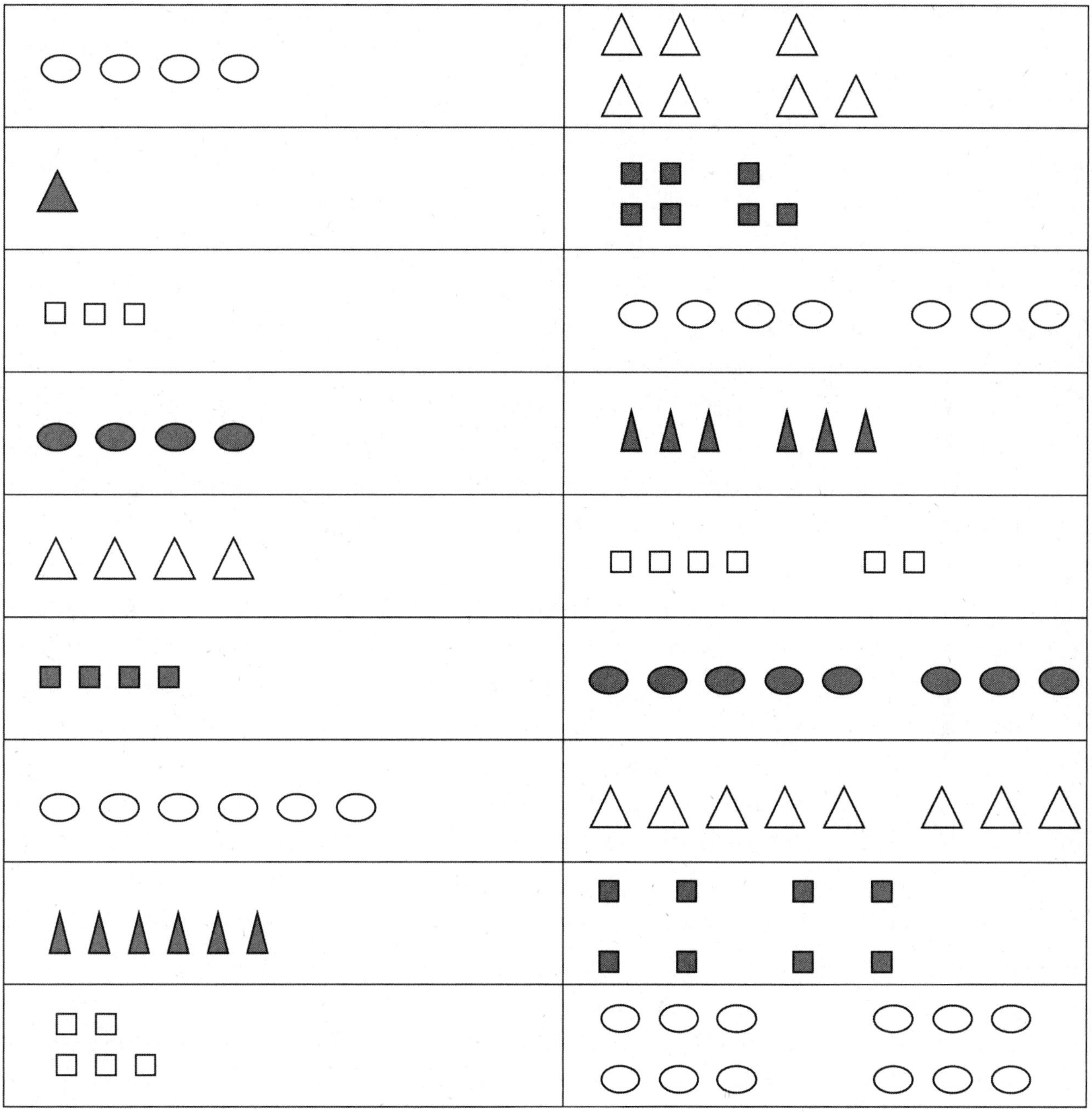

make 5

Count and Circle How Many

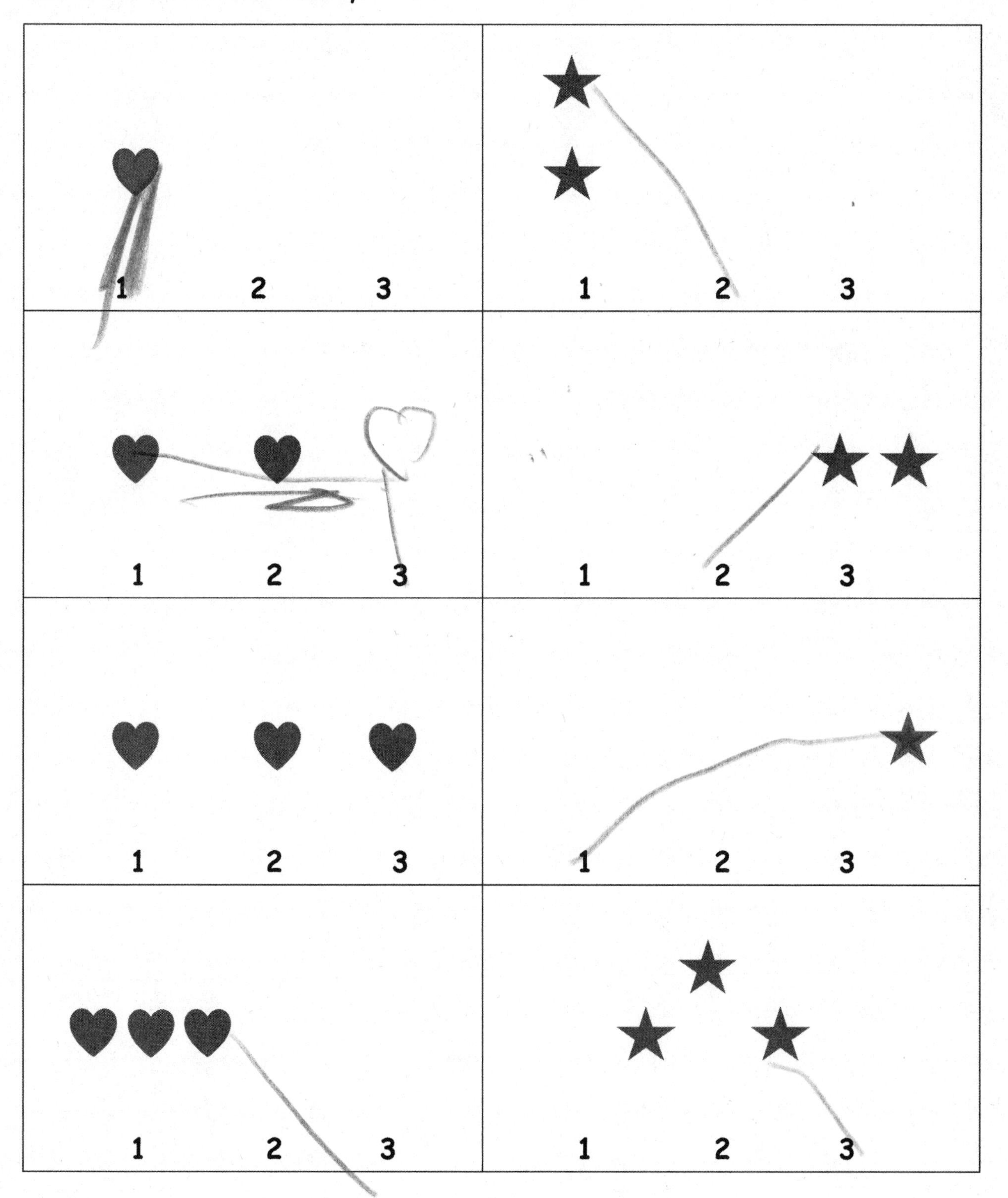

apple mat

Rekenrek to 5

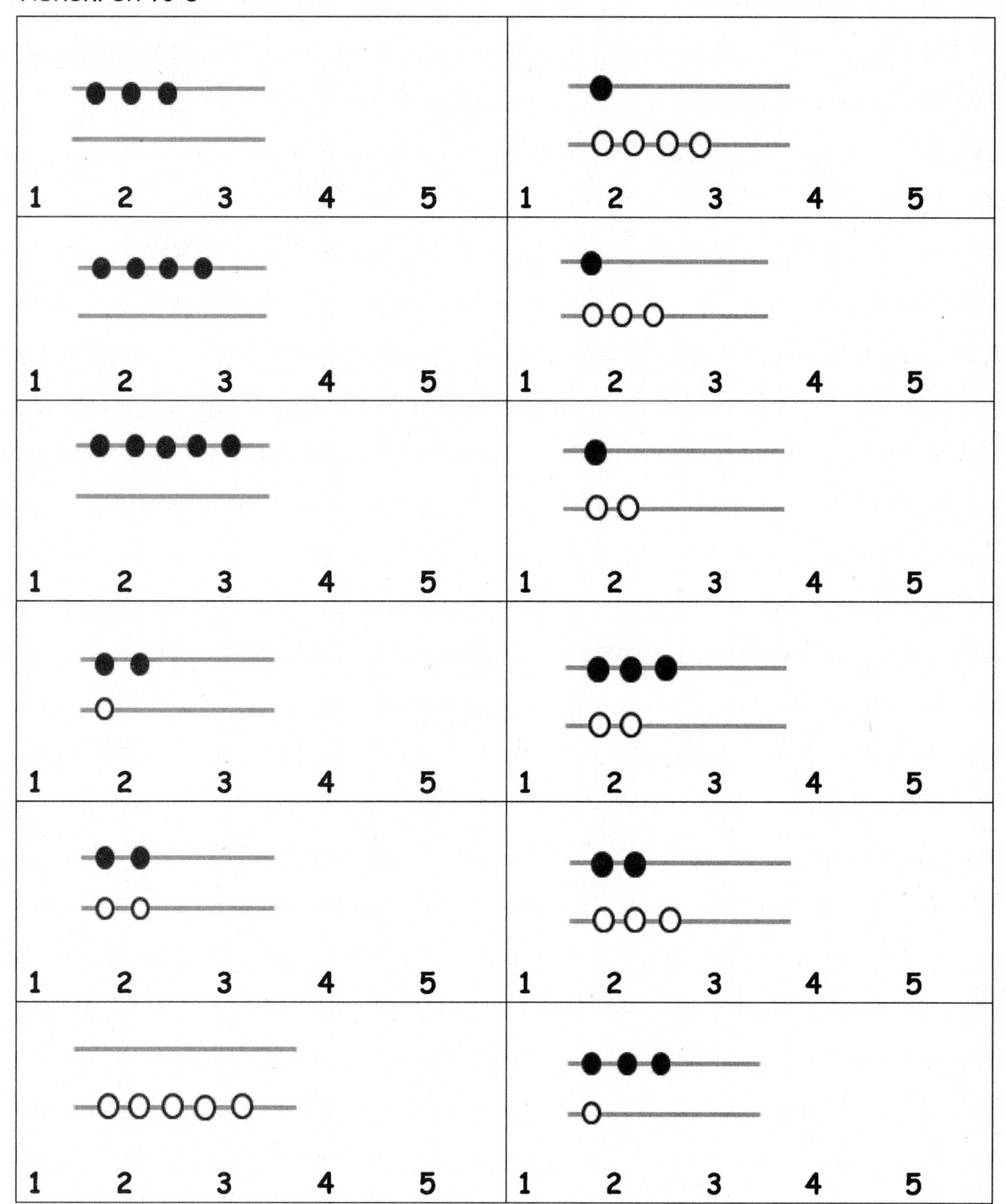

Grade K
Module 4

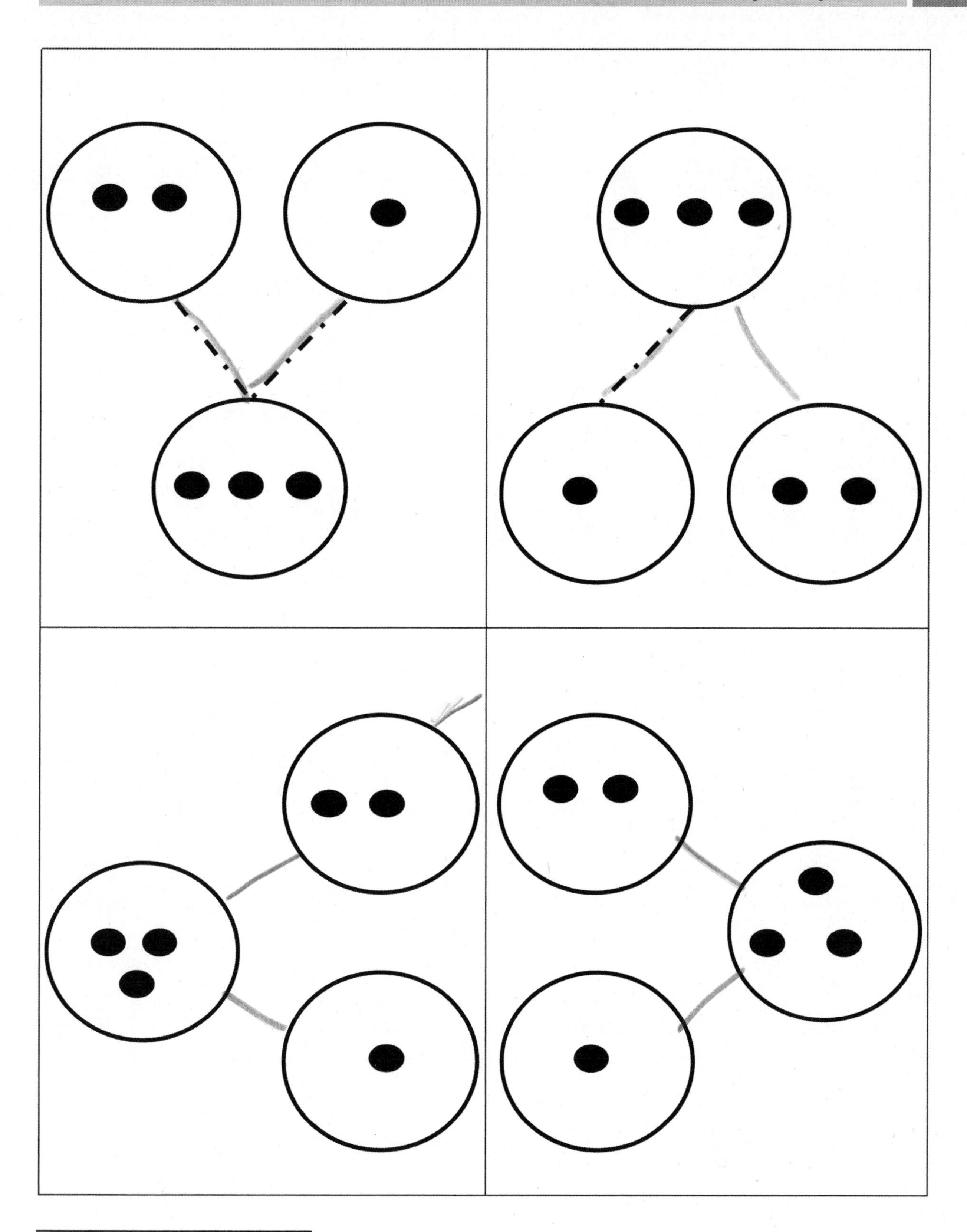

make a bond of 3

hidden numbers mat

Draw lines to make a bond of 4.

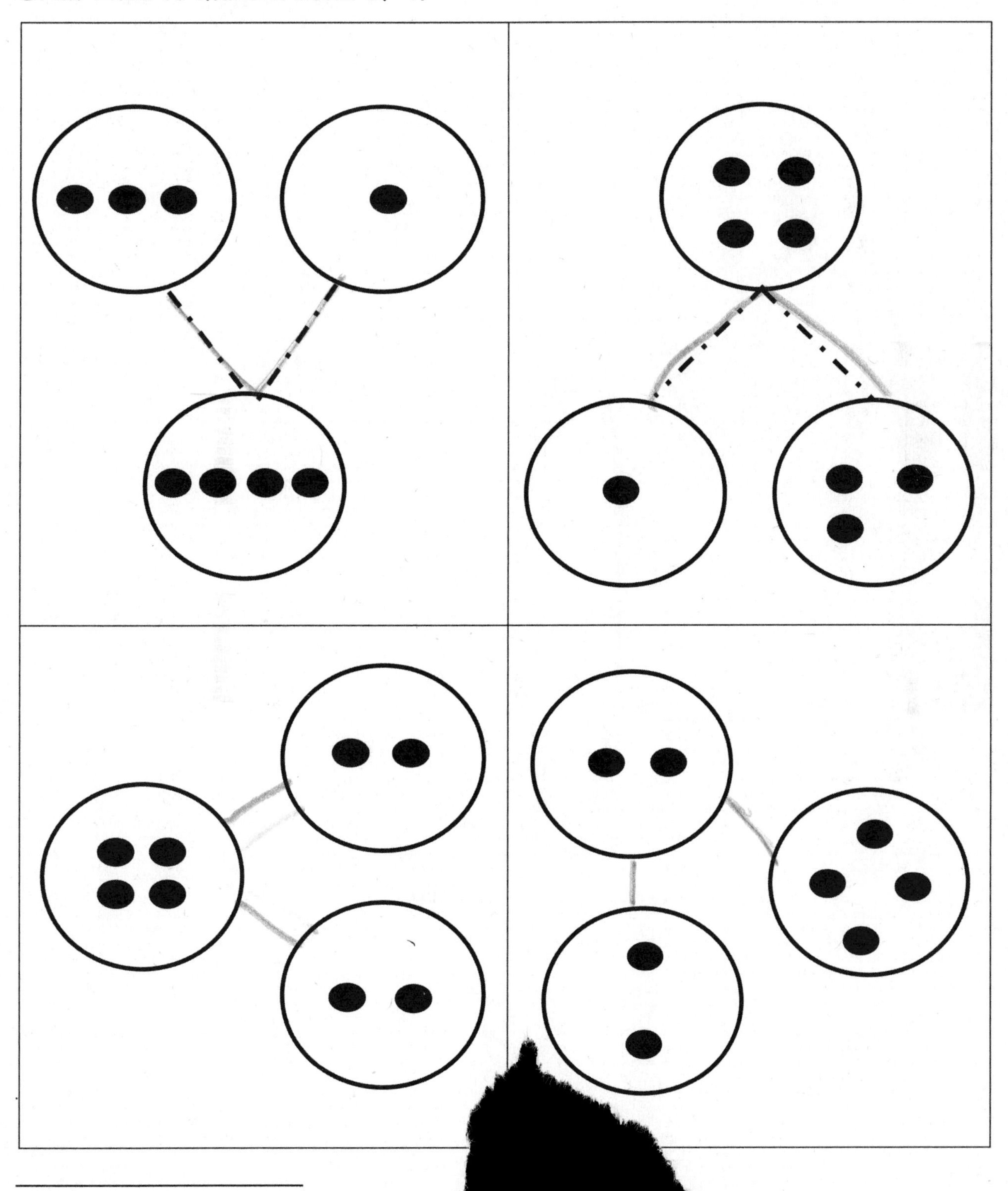

make a bond of 4

Draw lines to make a bond of 5.

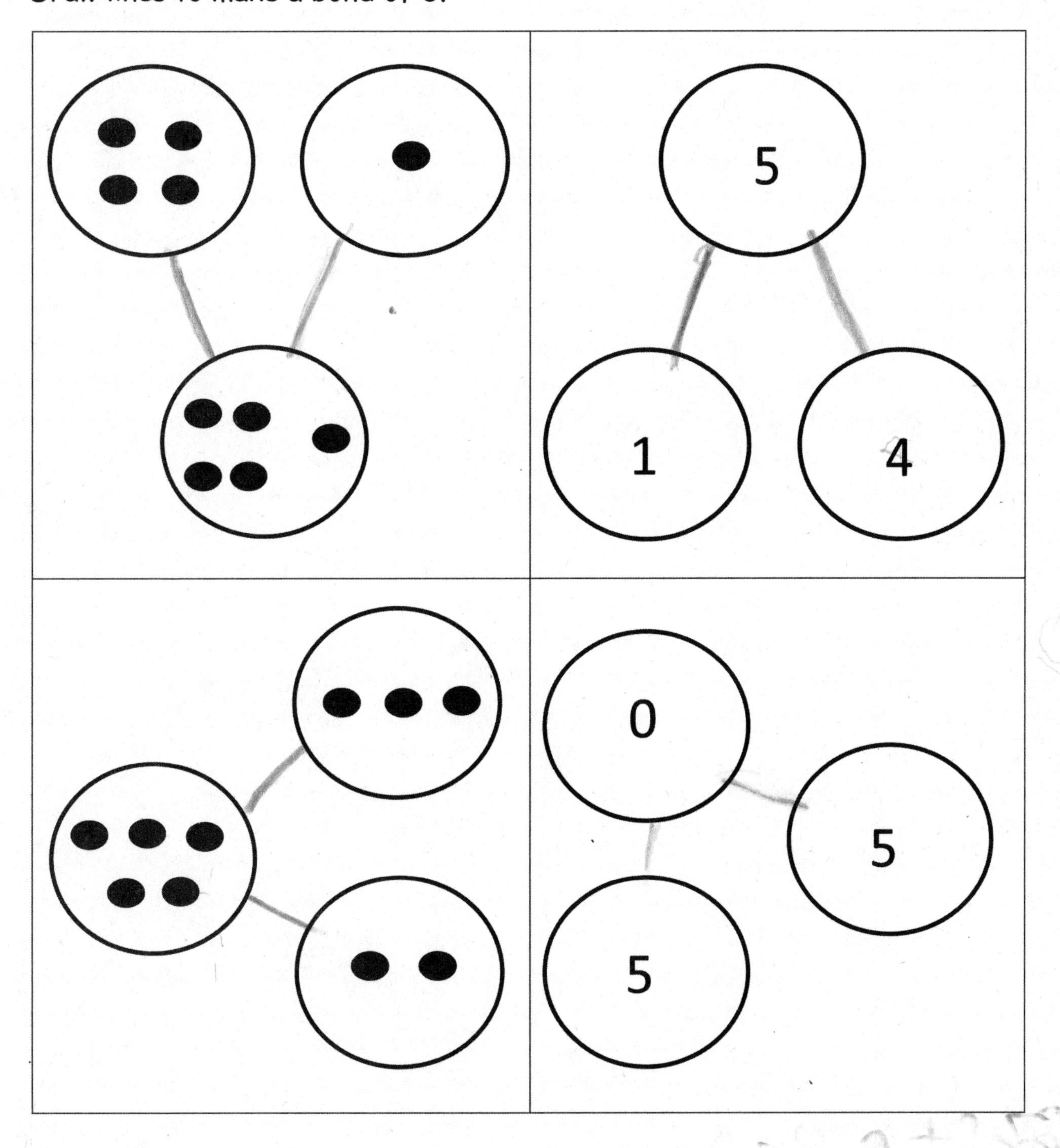

make a bond of 5

Circle the number needed to make 5.

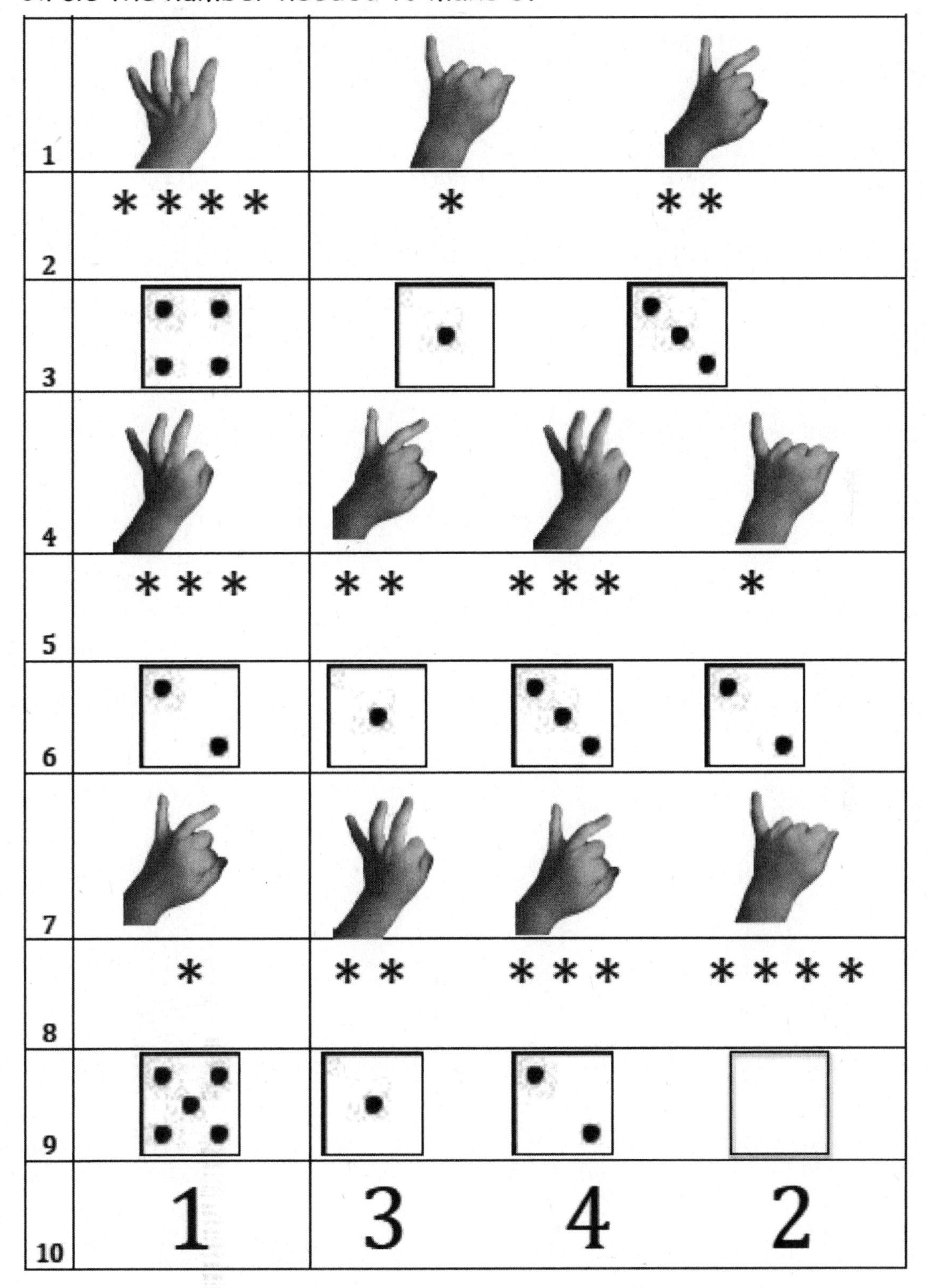

Circle the number to make 6.

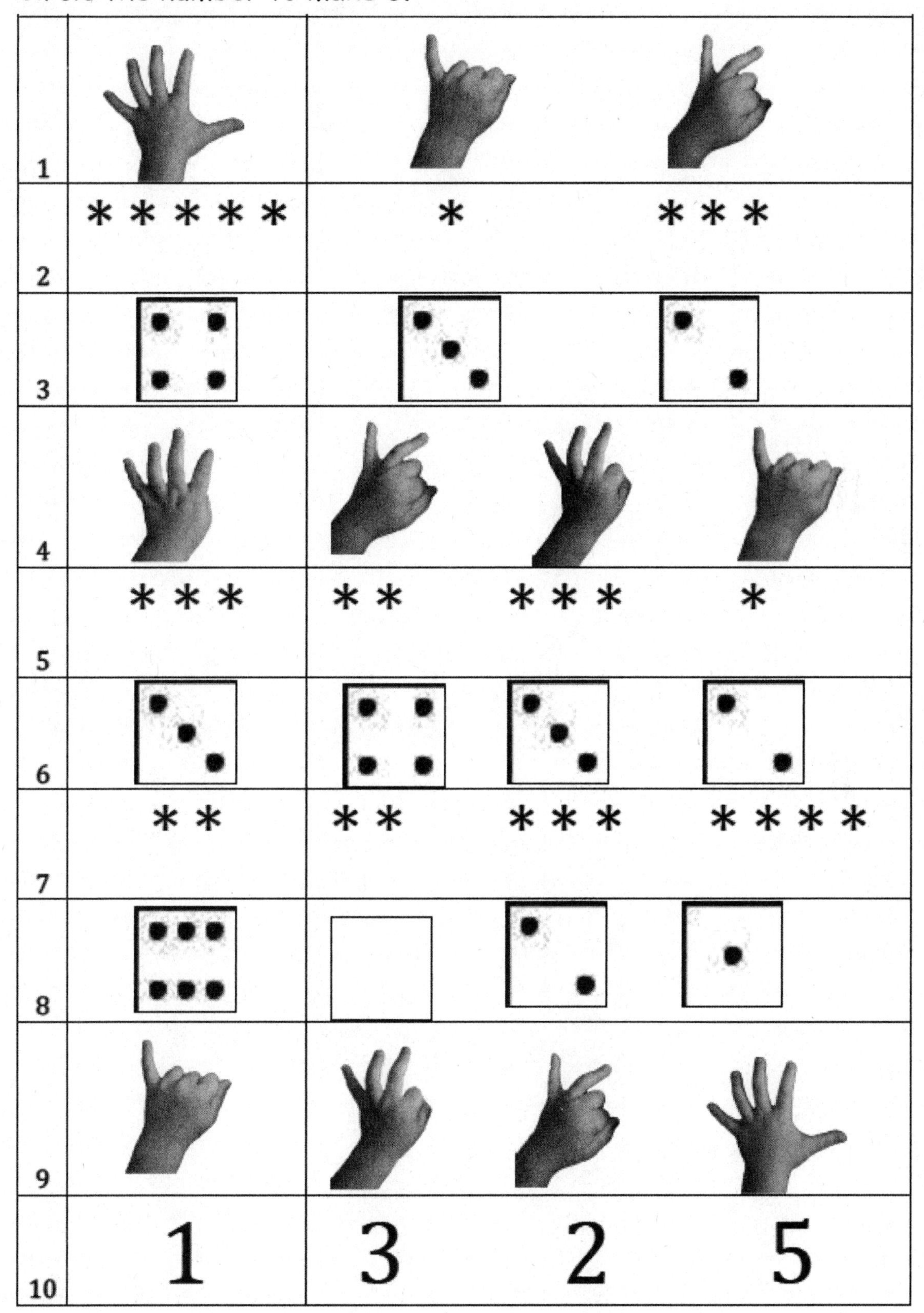

Draw more to make 5.

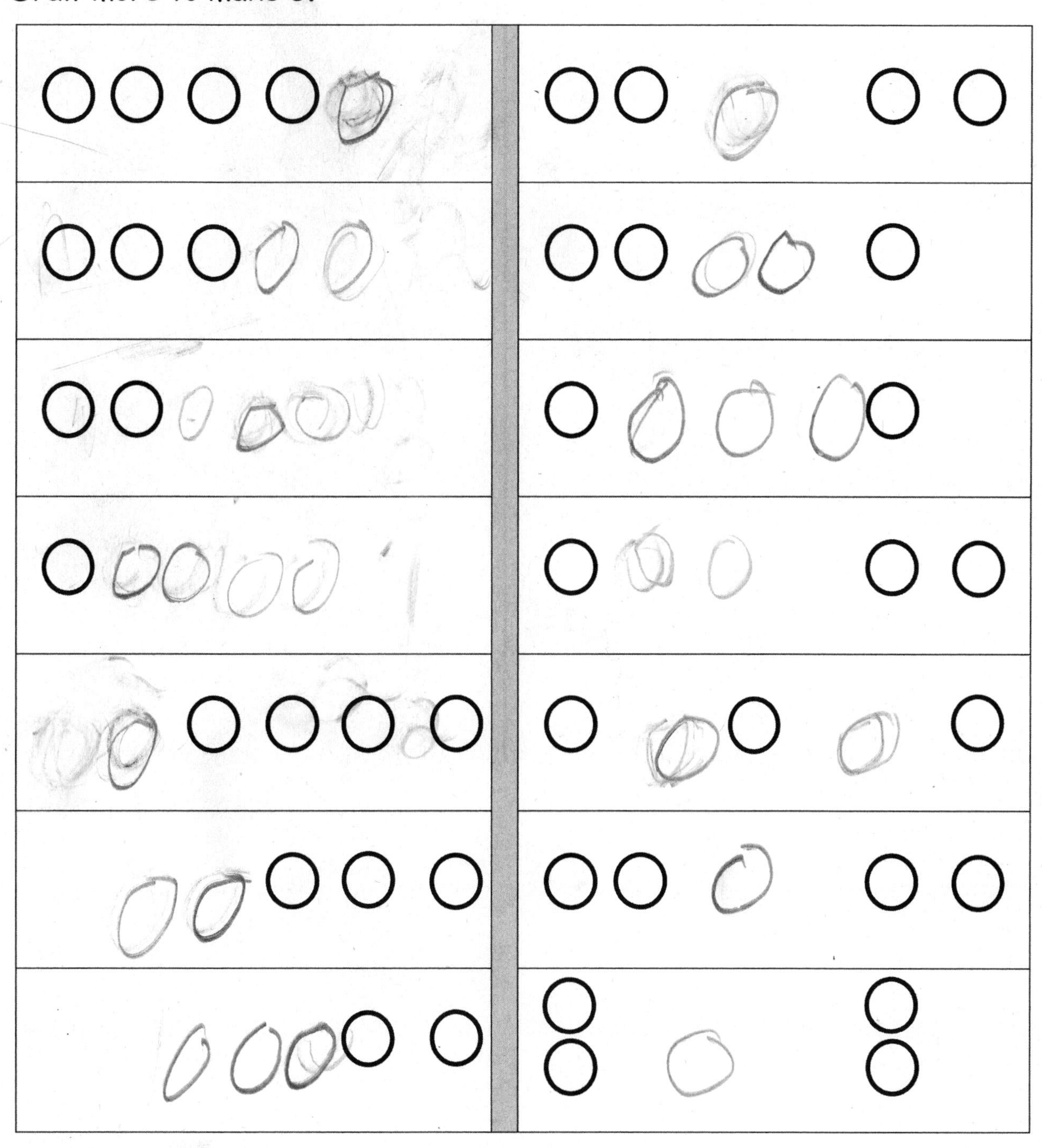

make 5

Circle the number to make 7.

1				
2				
3	* * * * *	* *	* * *	
	* * *	* * *	* * * *	* *
7				
8				
9	* *	* *	* * * * *	* * * *
10	2	2	5	4
11				
12				
13	1	2	6	5

Circle the number to make 8.

1		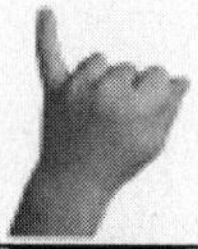		
2	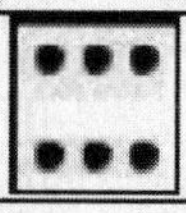	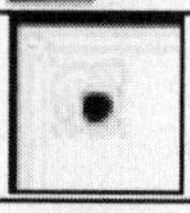	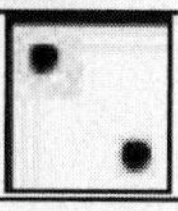	
3	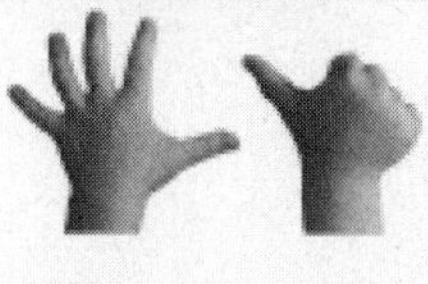			
4	* * * * *	* *	* * *	
5	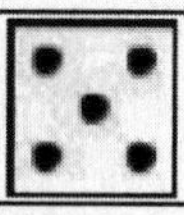	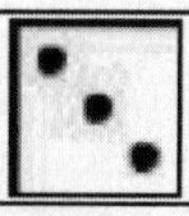	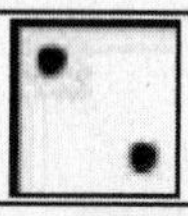	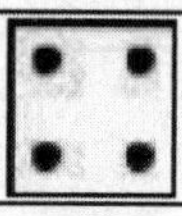
6				
7	* * * *	* * *	* * * *	* *
8	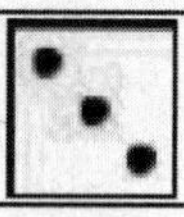	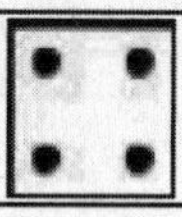	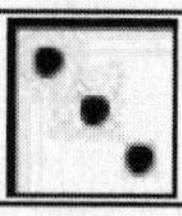	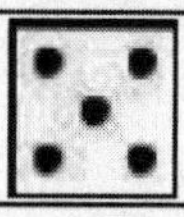
9			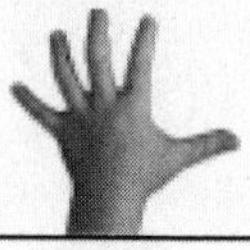	
10	* *	* * * * * *	* * * *	* * *
11	2	6	4	3
12			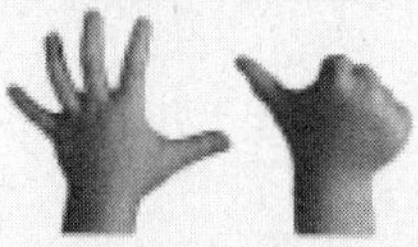	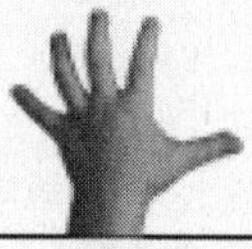
13	1	7	6	5

Circle the number to make 5.

1				
2	* * * *	* * *	*	
3	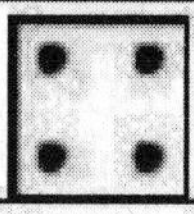	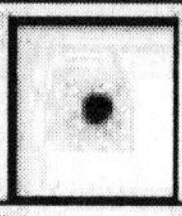	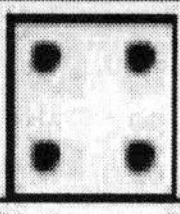	
4	4	1	4	
5		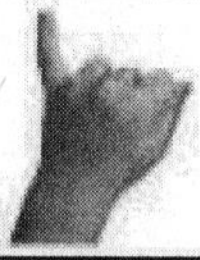		
6	* * *	* *	* * * *	*
7	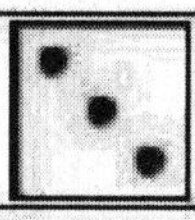	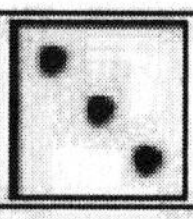	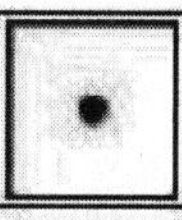	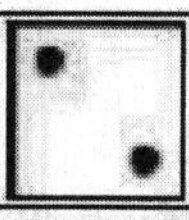
8	3	3	1	2
9				
10	* *	* * * *	* * *	* *
11	2	2	3	1
12	*	* * * * *	* * *	* * * *
13	1	4	5	3
14	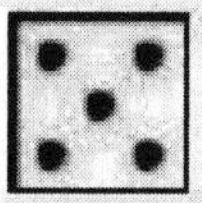	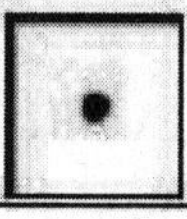	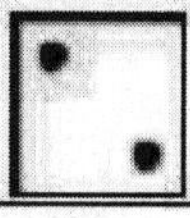	
15	5	2	1	0

Cross 1 out, and write how many.

EUREKA MATH

Lesson 20: Solve *take from with result unknown* expressions and equations using the minus sign with no unknown.

57

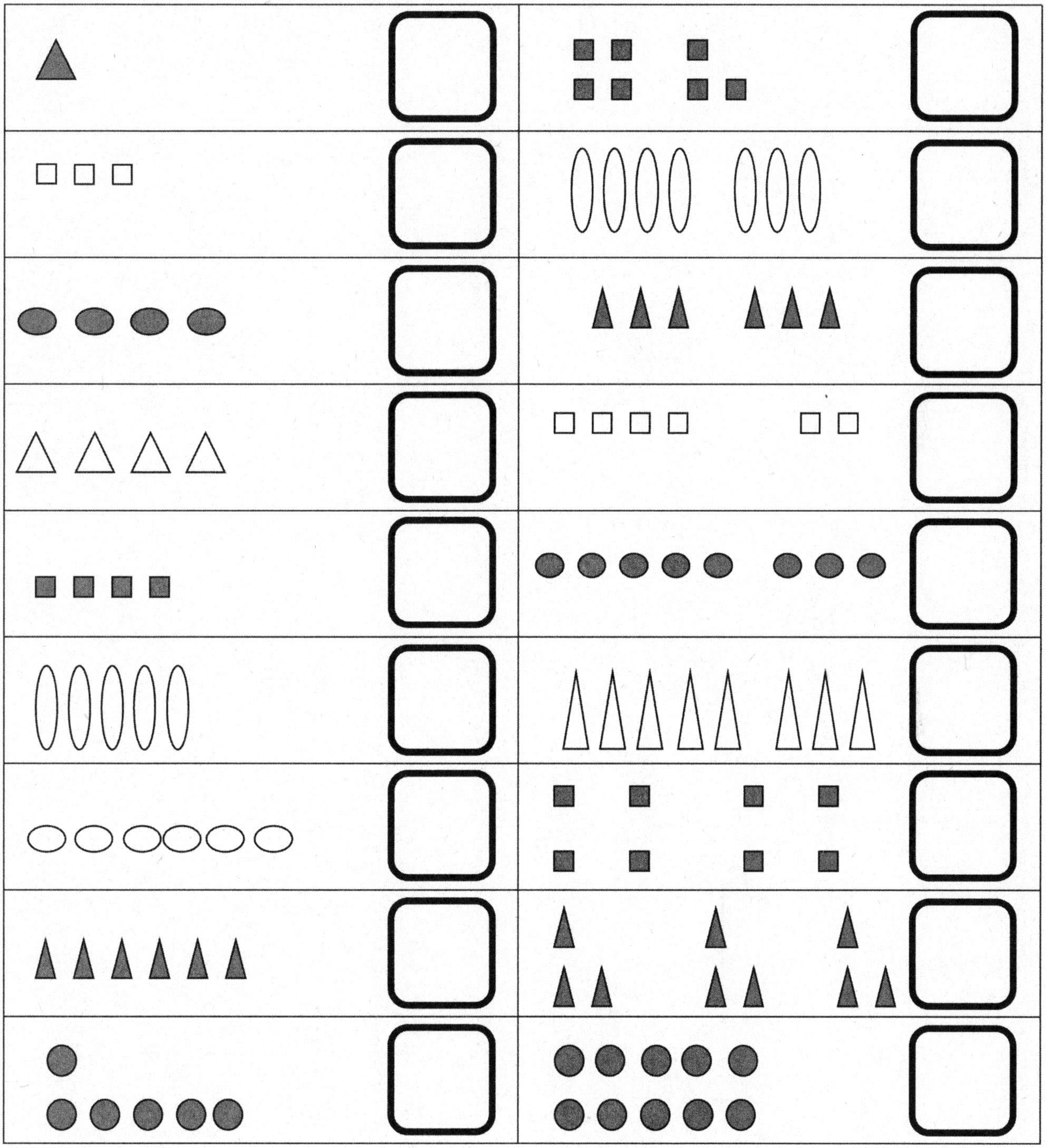

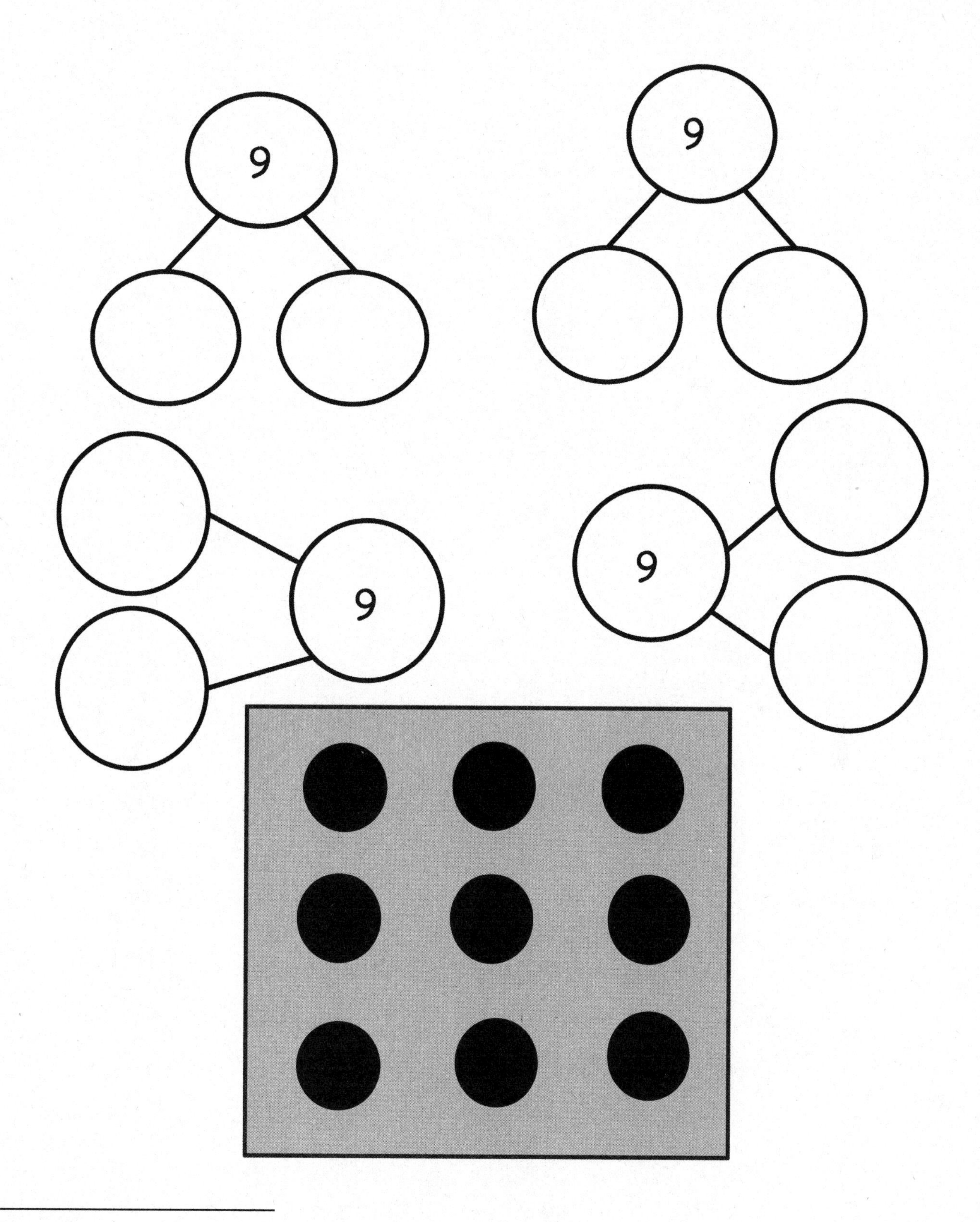

array of 9

EUREKA MATH

Lesson 25: Model decompositions of 9 using a story situation, objects, and number bonds.

63

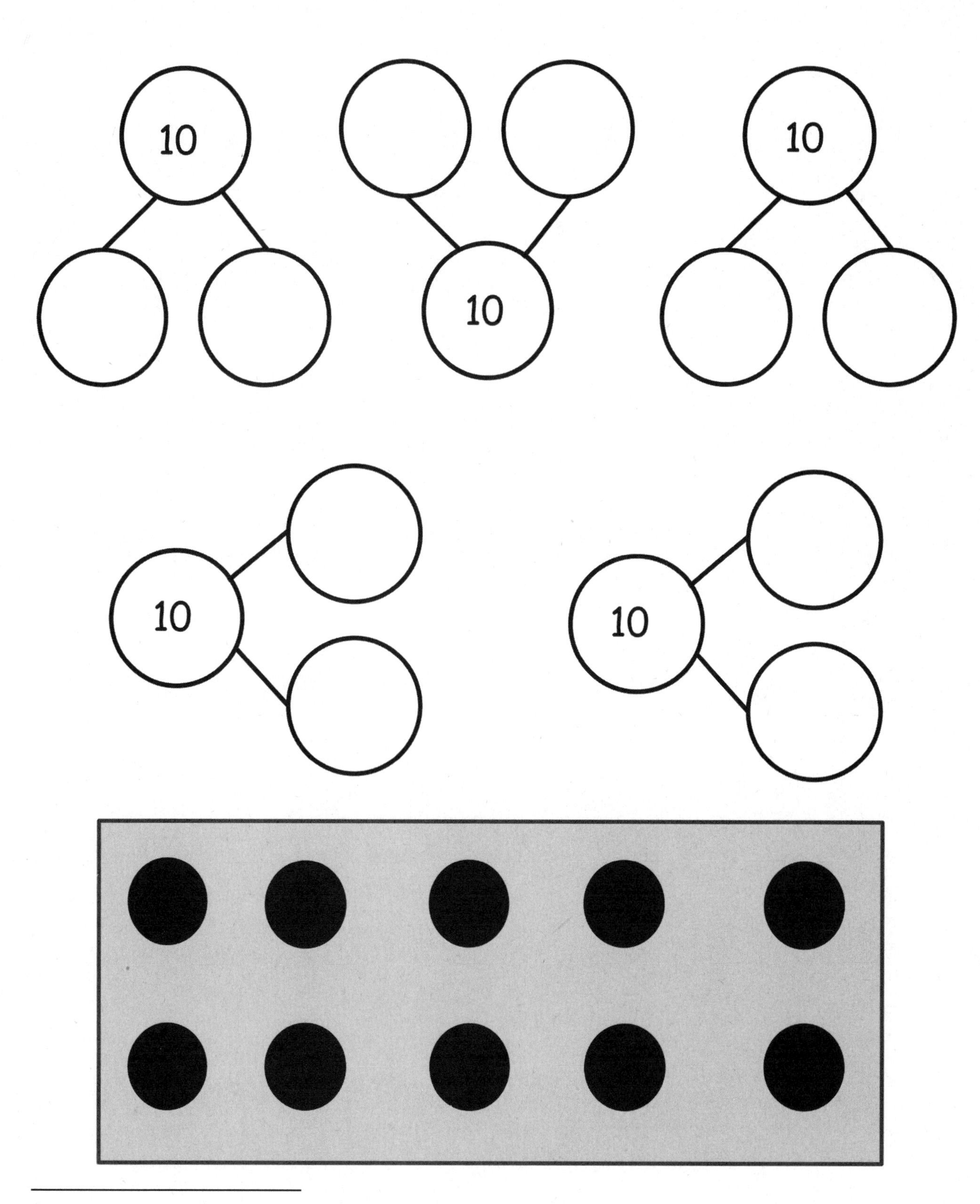

array of 10

Lesson 27: Model decompositions of 10 using a story situation, objects, and
number bonds.

65

Name ____________________ Date ____________

My Decomposition Practice

1 + 1 = ☐	2 = ☐ + ☐
☐ = 4 + 1	3 = ☐ + ☐
1 + 2 = ☐	2 + 2 = ☐
3 + 2 = ☐	☐ = 3 + 1
☐ = 1 + 3	3 = ☐ + ☐
2 + 1 = ☐	3 + 2 = ☐
1 + 4 = ☐	4 = ☐ + ☐
☐ = 3 + 2	4 = ☐ + ☐

Name ____________________ Date __________

My Subtraction Practice

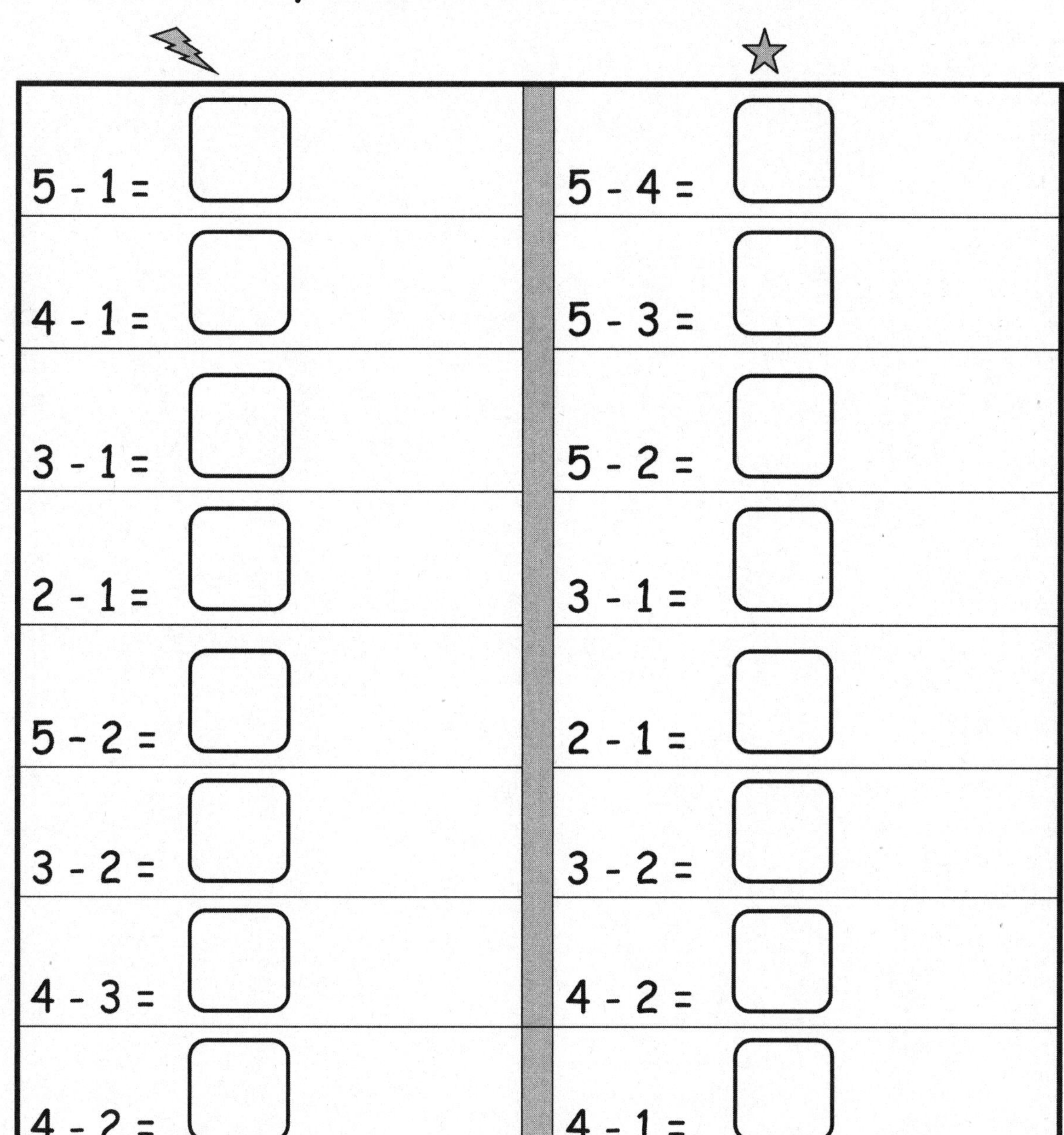

5 - 1 =	5 - 4 =
4 - 1 =	5 - 3 =
3 - 1 =	5 - 2 =
2 - 1 =	3 - 1 =
5 - 2 =	2 - 1 =
3 - 2 =	3 - 2 =
4 - 3 =	4 - 2 =
4 - 2 =	4 - 1 =

Lesson 29: Represent pictorial decomposition and composition addition stories to 9 with 5-group drawings and equations with no unknown.
71

Name ______________________________ Date ______________

My Subtraction Practice

5 - 1 = ☐	5 - 4 = ☐
☐ = 4 - 1	5 - 3 = ☐
3 - 1 = ☐	5 - 2 = ☐
2 - 1 = ☐	☐ = 3 - 1
☐ = 5 - 2	☐ = 2 - 1
3 - 2 = ☐	3 - 2 = ☐
4 - 3 = ☐	4 - 2 = ☐
☐ = 4 - 2	4 - 1 = ☐

Name ______________________________ Date ______________

My Mixed Practice to 5

1 + 1 = ☐	5 - 4 = ☐
☐ = 2 - 1	☐ = 2 + 3
3 + 1 = ☐	5 - 2 = ☐
4 - 1 = ☐	☐ = 3 - 1
☐ = 1 + 3	☐ = 2 + 1
3 + 2 = ☐	1 + 2 = ☐
5 - 3 = ☐	2 + 2 = ☐
☐ = 4 + 1	4 - 2 = ☐

Number Correct:

Name ______________________ Date ____________

Write the missing number.

1.	2 + 1 = ☐	11.	☐ = 3 + 2
2.	1 + 1 = ☐	12.	1 + 3 = ☐
3.	1 + 4 = ☐	13.	☐ = 2 + 2
4.	3 + 1 = ☐	14.	☐ = 1 + 2
5.	2 + 2 = ☐	15.	1 + 4 = ☐
6.	2 + 3 = ☐	16.	☐ = 2 + 3
7.	1 + 2 = ☐	17.	☐ = 5 – 1
8.	4 + 1 = ☐	18.	5 – 2 = ☐
9.	3 + 2 = ☐	19.	1 + 0 = ☐
10.	1 + 3 = ☐	20.	5 + 0 = ☐

Number Correct:

Name ______________________ Date ____________

Write the missing number.

1.	2 - 1 = ☐	11.	☐ = 4 - 2
2.	4 - 1 = ☐	12.	5 - 3 = ☐
3.	5 - 1 = ☐	13.	☐ = 3 - 1
4.	3 - 1 = ☐	14.	☐ = 5 - 2
5.	3 - 2 = ☐	15.	4 - 1 = ☐
6.	4 - 2 = ☐	16.	☐ = 5 - 4
7.	5 - 3 = ☐	17.	☐ = 5 - 1
8.	5 - 2 = ☐	18.	5 - 1 = ☐
9.	4 - 3 = ☐	19.	1 - 0 = ☐
10.	5 - 4 = ☐	20.	5 - 5 = ☐

Number Correct:

Name ______________________ Date ____________

Write the missing number.

1.	2 + 1 = ☐	11.	☐ = 1 + 2
2.	4 + 1 = ☐	12.	5 + 0 = ☐
3.	5 - 1 = ☐	13.	☐ = 3 - 1
4.	3 + 1 = ☐	14.	☐ = 2 + 2
5.	3 + 2 = ☐	15.	4 - 1 = ☐
6.	4 - 2 = ☐	16.	☐ = 5 - 4
7.	5 - 3 = ☐	17.	☐ = 5 - 1
8.	5 - 2 = ☐	18.	3 + 0 = ☐
9.	2 + 3 = ☐	19.	1 - 0 = ☐
10.	5 - 4 = ☐	20.	5 - 5 = ☐

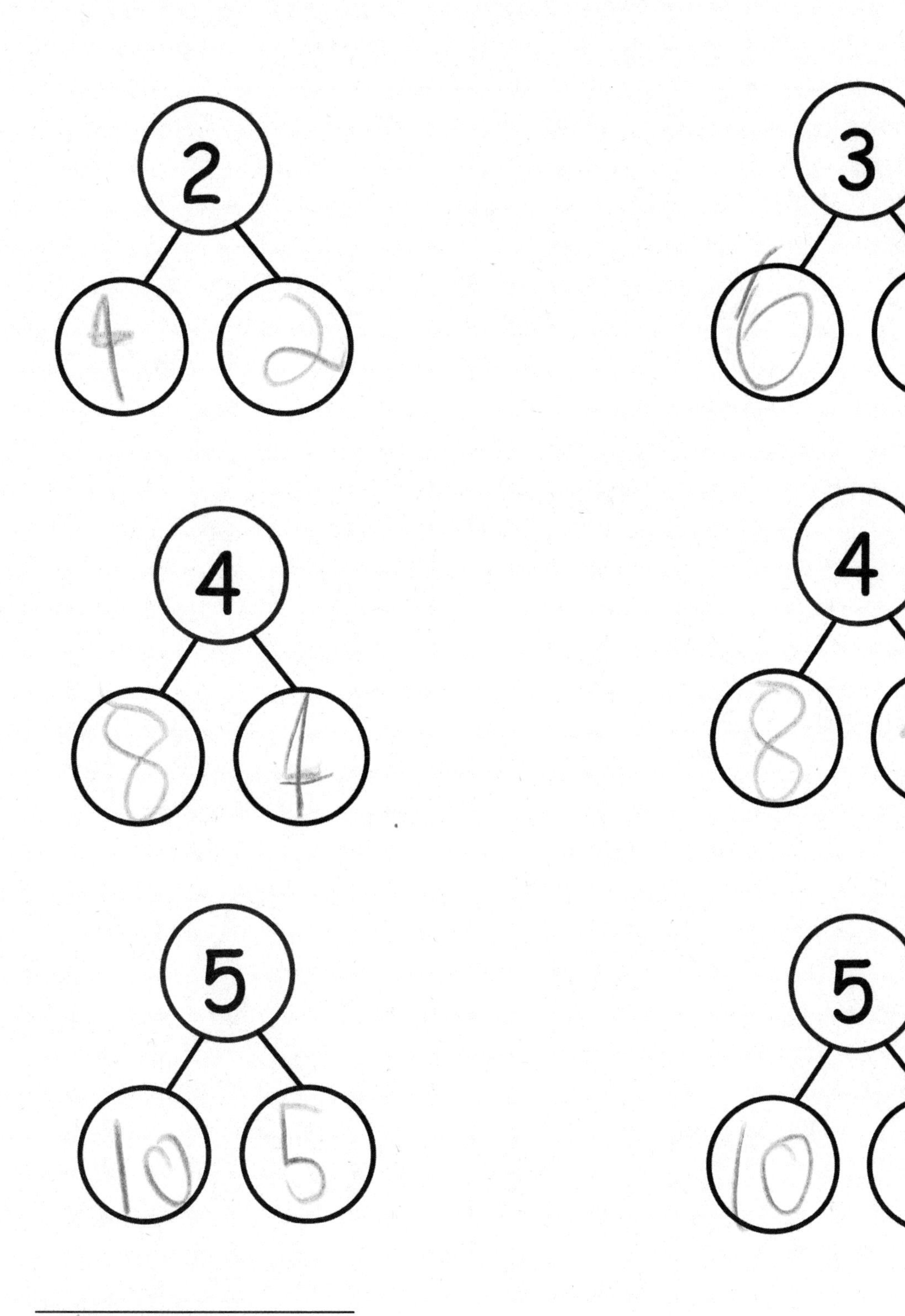

break apart numbers

Imagine more to add to 5, and write the addition sentence in the box.

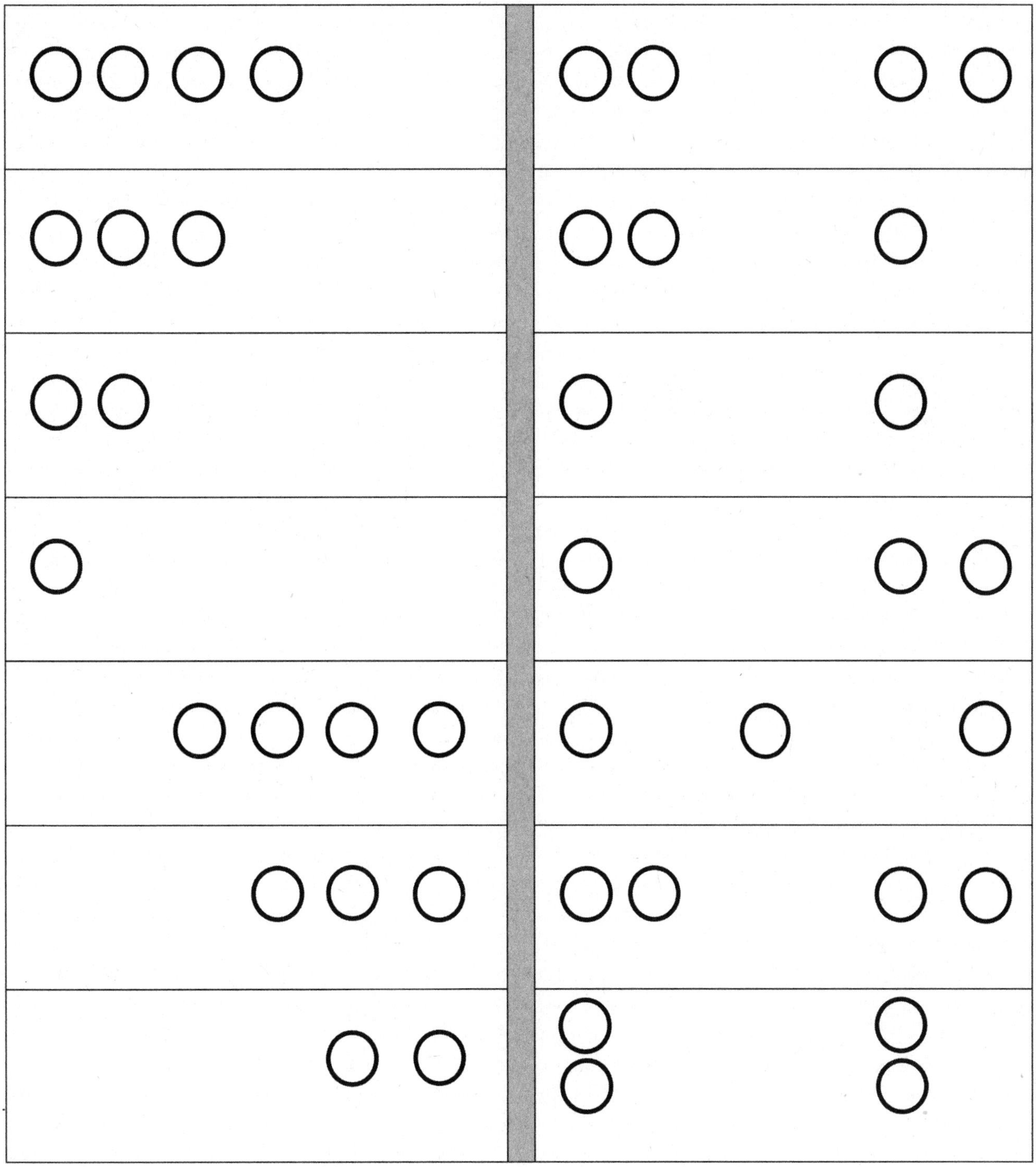

imagine more to add to 5

Cross out 2, and finish the subtraction sentence.

★ ★ ★	3 - 2 = ____
☾ ☾ ☾ ☾	4 - 2 = ____
★ ★ ★ ★ ★	5 - 2 = ____
☾ ☾	2 - 2 = ____
★ ★ ★ ★	4 - ____ = ____
☾ ☾ ☾ ☾ ☾	5 - ____ = ____

cross out 2

apple tree

draw more to make 10

Grade K
Module 5

Name ______________________________ Date ____________________

Circle 10.

circle 10

Name ______________________________ Date ____________________

Circle sets of 10, and tell how many.

____

circle 10 ones

Name ______________________________ Date ______________________

Count the objects in each group and write the number.

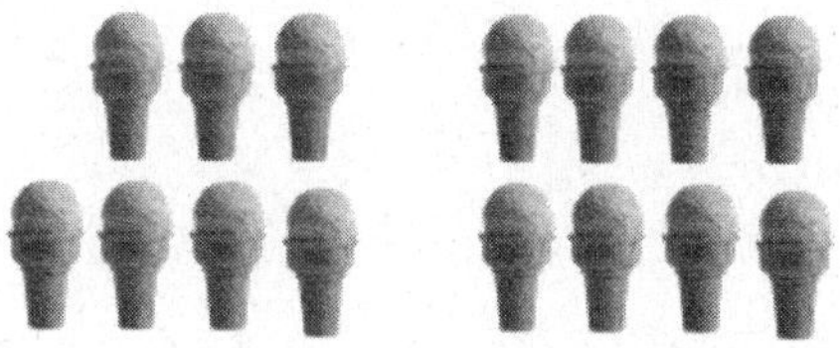

teen counting array

Name ______________________ Date ________________

Whisper count and draw in more shapes to match the number.

14

12

15

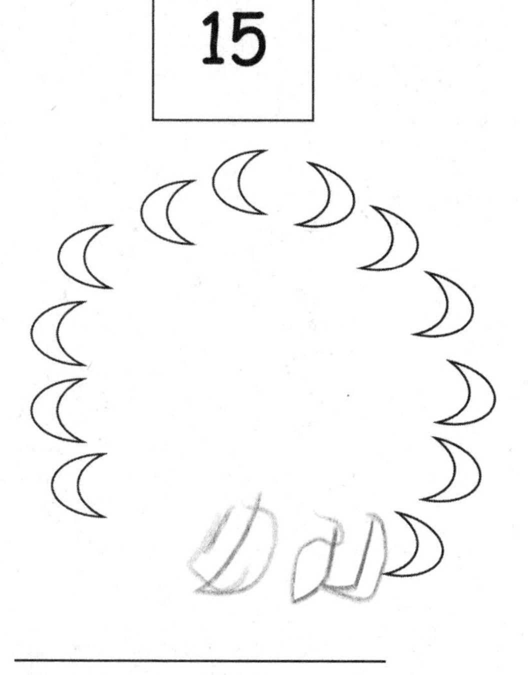

17

teen circular-counting

Whisper count and draw in more shapes to match the number.

16

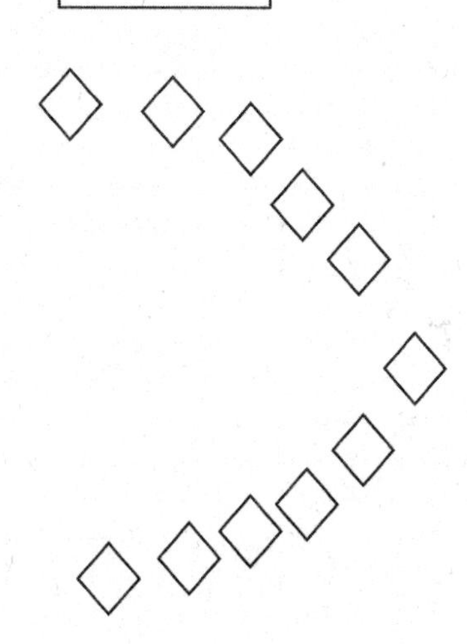

19

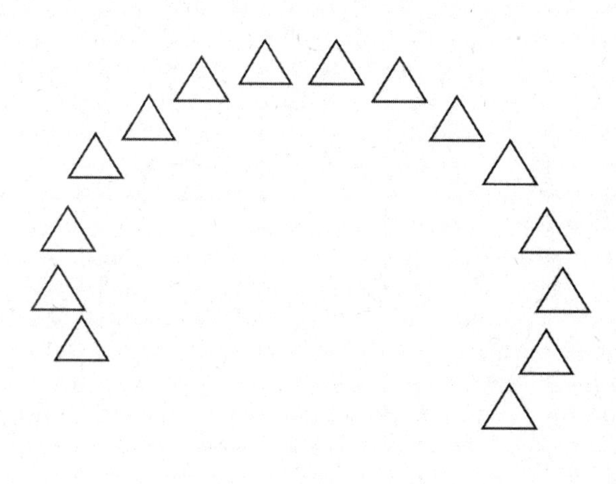

13

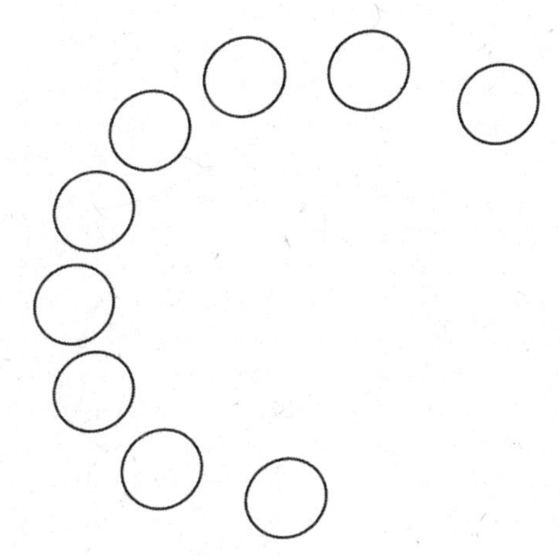

20

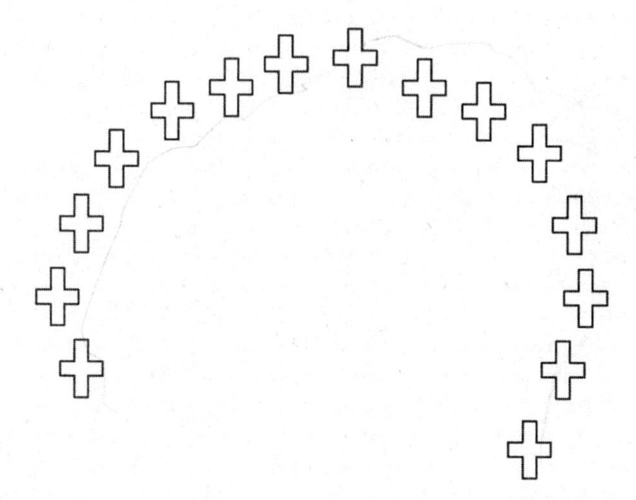

teen circular-counting

Rekenrek

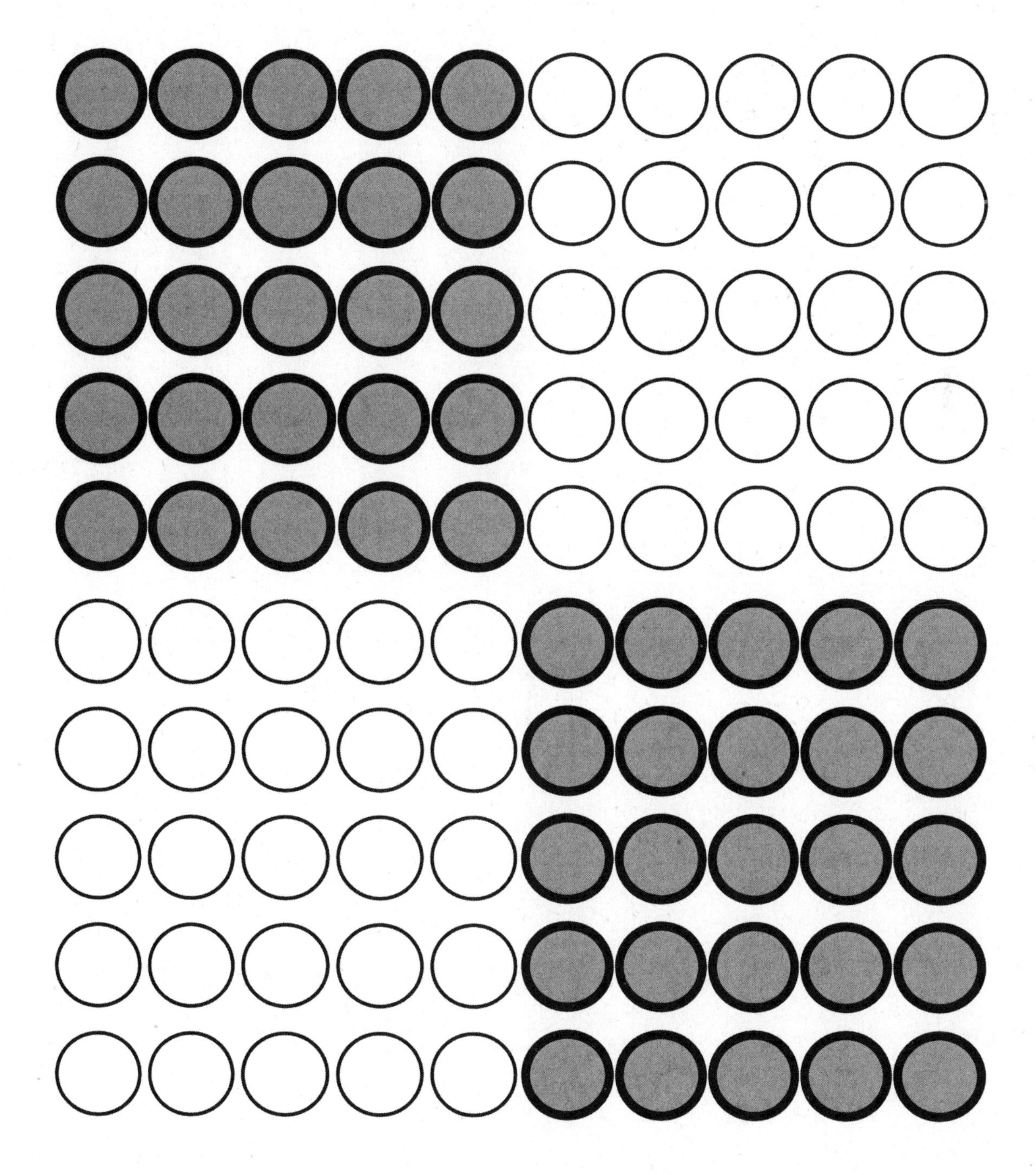

Rekenrek dot paper

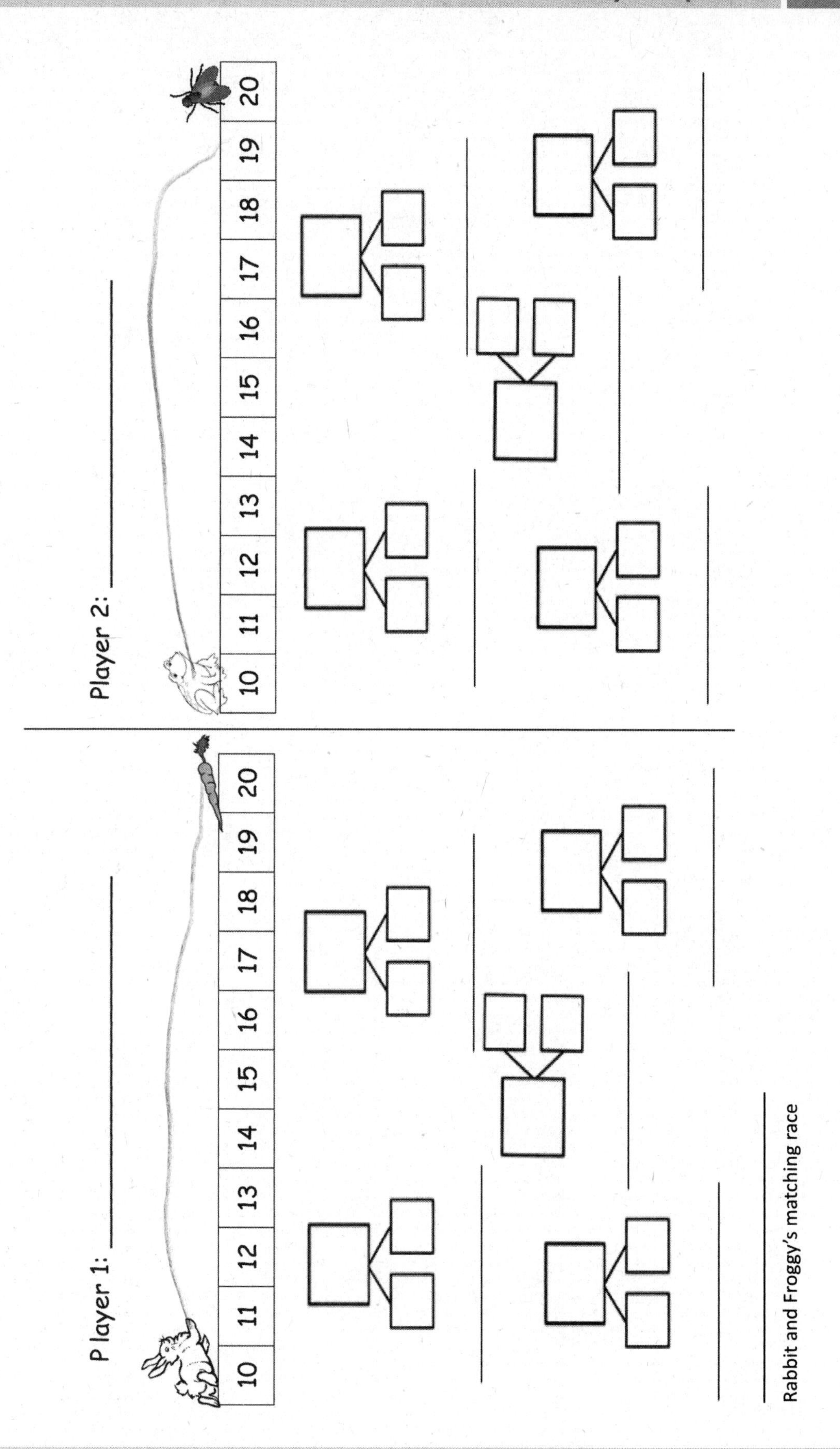
Player 1: ____________
10
11
12
13
14
15
16
17
18
19
20
Player 2: ____________
10
11
12
13
14
15
16
17
18
19
20
Rabbit and Froggy's matching race

Grade K
Module 6

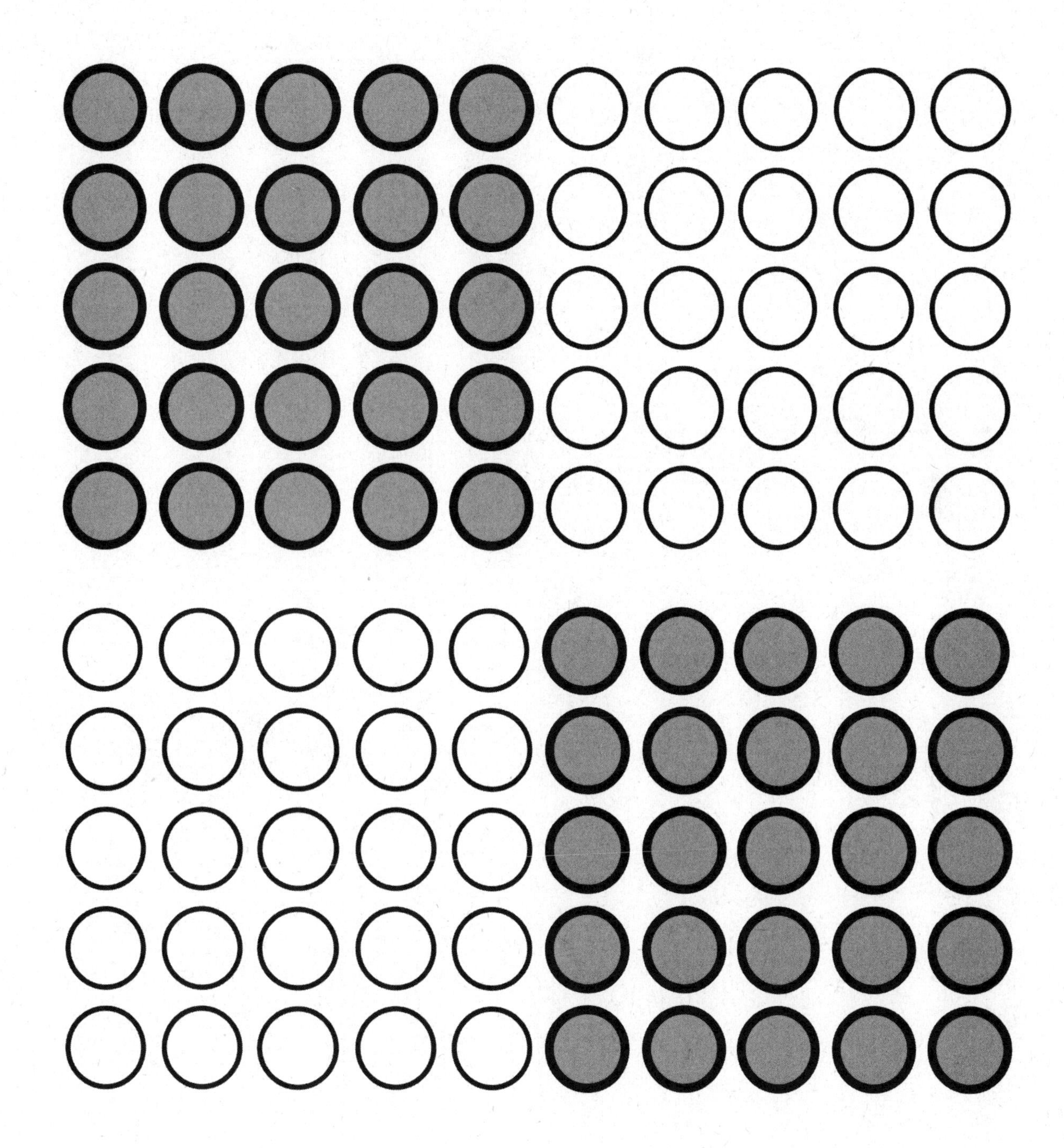

Rekenrek dot paper

Number Corrrect:

Name ____________________ Date ____________

Write the missing number.

1.	2 + 1 = ☐	11.	☐ = 3 + 2
2.	1 + 1 = ☐	12.	1 + 3 = ☐
3.	1 + 4 = ☐	13.	☐ = 2 + 2
4.	3 + 1 = ☐	14.	☐ = 1 + 2
5.	2 + 2 = ☐	15.	1 + 4 = ☐
6.	2 + 3 = ☐	16.	☐ = 2 + 3
7.	1 + 2 = ☐	17.	☐ = 5 − 1
8.	4 + 1 = ☐	18.	5 − 2 = ☐
9.	3 + 2 = ☐	19.	1 + 0 = ☐
10.	1 + 3 = ☐	20.	5 + 0 = ☐

Number Correct:

Name ____________________ Date ____________

Write the missing number.

1.	2 – 1 = ☐	11.	☐ = 4 – 2
2.	4 – 1 = ☐	12.	5 – 3 = ☐
3.	5 – 1 = ☐	13.	☐ = 3 – 1
4.	3 – 1 = ☐	14.	☐ = 5 – 2
5.	3 – 2 = ☐	15.	4 – 1 = ☐
6.	4 – 2 = ☐	16.	☐ = 5 – 4
7.	5 – 3 = ☐	17.	☐ = 5 – 1
8.	5 – 2 = ☐	18.	5 – 1 = ☐
9.	4 – 3 = ☐	19.	1 – 0 = ☐
10.	5 – 4 = ☐	20.	5 – 5 = ☐

Number Correct:

Name ____________________ Date ____________

Write the missing number.

1.	2 + 1 = ☐	11.	3 + 2 = ☐
2.	2 - 1 = ☐	12.	3 - 2 = ☐
3.	3 + 1 = ☐	13.	4 + 0 = ☐
4.	3 - 1 = ☐	14.	4 - 0 = ☐
5.	4 + 1 = ☐	15.	5 + 0 = ☐
6.	4 - 1 = ☐	16.	5 - 0 = ☐
7.	1 + 1 = ☐	17.	5 - 5 = ☐
8.	1 - 1 = ☐	18.	4 + 1 = ☐
9.	2 + 2 = ☐	19.	5 - 4 = ☐
10.	2 - 2 = ☐	20.	5 - 1 = ☐

Number Correct:

Name ______________________ Date ____________

Write the missing number.

1.	2 + 1 = ☐	11.	☐ = 1 + 2
2.	4 + 1 = ☐	12.	5 + 0 = ☐
3.	5 - 1 = ☐	13.	☐ = 3 - 1
4.	3 + 1 = ☐	14.	☐ = 2 + 2
5.	3 + 2 = ☐	15.	4 - 1 = ☐
6.	4 - 2 = ☐	16.	☐ = 5 - 4
7.	5 - 3 = ☐	17.	☐ = 5 - 1
8.	5 - 2 = ☐	18.	3 + 0 = ☐
9.	2 + 3 = ☐	19.	1 - 0 = ☐
10.	5 - 4 = ☐	20.	5 - 5 = ☐

Name ______________________ Date ____________

Add. Color the blocks using the code for the total.

1—RED	2—ORANGE	3—YELLOW
4—GREEN	5—BLUE	

0 + 1	1 + 1	2 + 1	3 + 1	4 + 1
0 + 2	1 + 2	2 + 2	3 + 2	
0 + 3	1 + 3	2 + 3		
0 + 4	1 + 4			
0 + 5				

color by answer addition

Name ______________________________ Date ______________

Subtract. Color the blocks using the code for the difference.

0—PURPLE 1—RED 2—ORANGE 3—YELLOW
4—GREEN 5—BLUE

1 - 0	2 - 0	3 - 0	4 - 0	5 - 0
1 - 1	2 - 1	3 - 1	4 - 1	5 - 1
	2 - 2	3 - 2	4 - 2	5 - 2
		3 - 3	4 - 3	5 - 3
			4 - 4	5 - 4
				5 - 5

color by answer subtraction

Number Correct:

Name ______________________ Date __________

Write the number of dots needed to make 10 dots.

1.	●●●●● ●●●●	9.	●
2.	●●●●● ●●●	10.	●●●●● ●●●●
3.	●●●●● ●●	11.	●●●●● ●●●
4.	●●●●● ●	12.	●●
5.	●●●●●	13.	●●●●● ●●
6.	●●●●	14.	●●●
7.	●●●	15.	●●●●● ●
8.	●●	16.	●●●●

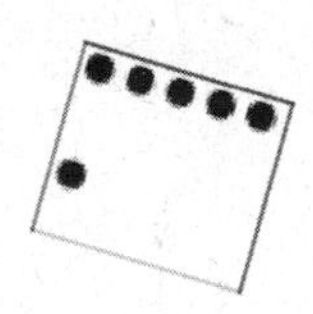

1	2	3	4	5	6	7	8	9	10

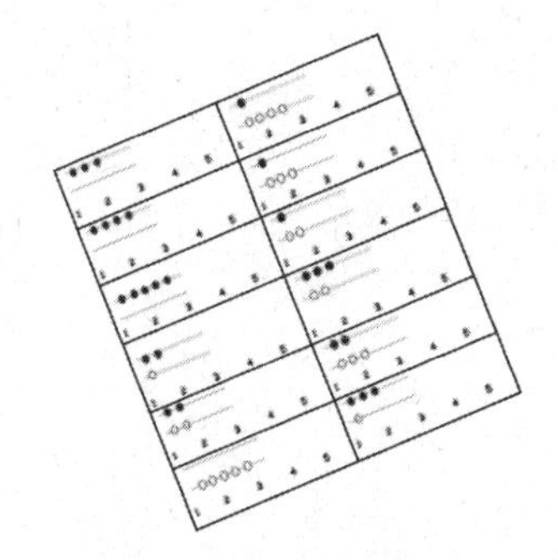

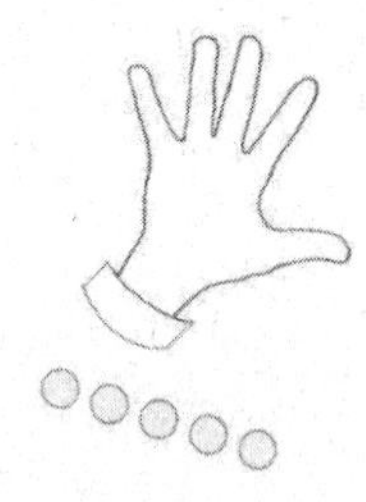

I'm Getting Ready for First Grade!

My Math Fluency Kit

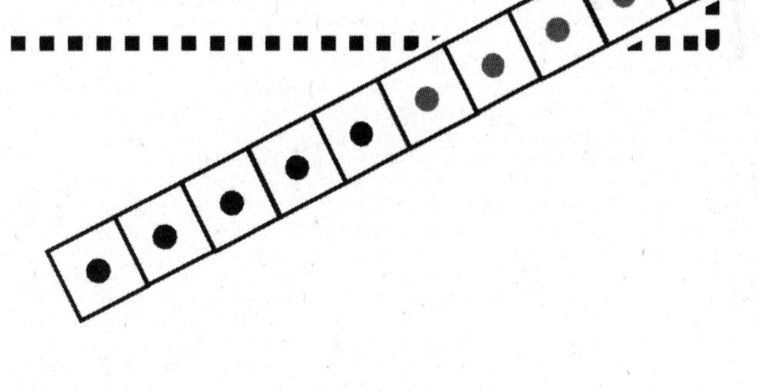

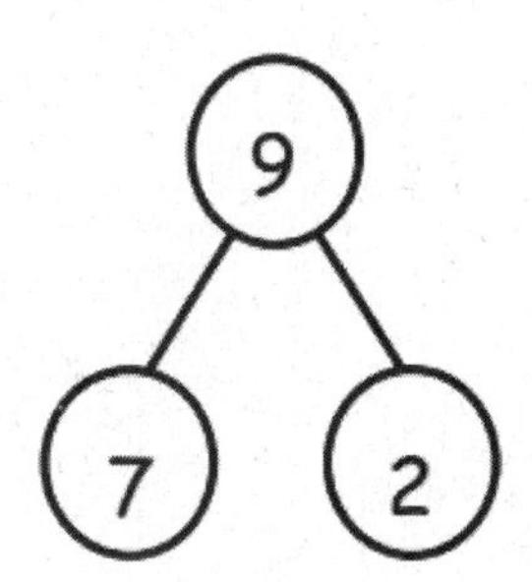

Name ____________________

fluency kit

Name ______________________________ Date ____________

My Plan to Get Ready for First Grade Math.

This is a picture of someone who can help me practice.

This is a picture of where I will practice.

This is ME getting ready for first grade!

---

fluency kit

Name ___________________________

My Sprint Progress Log

Practice your number sentences and Sprints on your personal white board. Ask an adult to time you. Keep track of how you improve over the summer.

Date	Time

Are you getting better at your number sentences?

fluency kit

Credits

Great Minds® has made every effort to obtain permission for the reprinting of all copyrighted material. If any owner of copyrighted material is not acknowledged herein, please contact Great Minds for proper acknowledgment in all future editions and reprints of this module.